The Leadership Journey Of Kelly Ortberg

Resilience, Innovation, And Transformation In The Aerospace Industry.

Pauline R. Valerie

Copyright © 2024 Pauline R. Valerie

All rights reserved. No part of this publication may be reproduced, distributed, or transmitted in any form or by any means, including photocopying, recording, or other electronic or mechanical methods, without the prior written permission of the publisher, except in the case of brief quotations embodied in critical reviews and certain other noncommercial uses permitted by copyright law.

Table Of Contents

Introduction

Chapter One; Early Life and Education

Chapter Two; Path to Leadership

Chapter Three; The Rise to CEO of Boeing

Chapter Four; Challenges of a New CEO

Chapter Five; The Boeing 737 Max Crisis

Chapter Six; Restoring Boeing's Reputation

Chapter Seven; Financial Struggles and Debt Crisis

Chapter Eight; The 2023 Machinist Strike

Chapter Nine; Labor Relations and Union Negotiations

Chapter Ten; Handling a Workforce in Rebellion
Chapter Eleven; Economic Impact of the Strike
Chapter Twelve; Furloughs and Cost-Cutting Measures
Chapter Thirteen; The Road to Resolution
Chapter Fourteen; The Future of Boeing Under Ortberg
Chapter Fifteen; Legacy in Aerospace Leadership
Chapter Sixteen; Personal Life and Philosophy
Conclusion

Introduction

Transformative. That's the word that encapsulates the journey of Kelly Ortberg, a leader who has not only navigated the turbulent skies of the aerospace industry but has also reshaped the very fabric of Boeing.

In an era defined by rapid technological advancements and unprecedented challenges, Kelly emerged as a guiding force, steering the company through crises and towards a future of innovation and resilience. From his early days as a curious child fascinated by machinery to his ascent as the CEO of one of the world's most iconic aerospace companies, Kelly's story is a compelling narrative of ambition, integrity, and unwavering commitment.

He embodies the idea that true leadership transcends titles and corporate strategies; it is about people, relationships, and the values that underpin our actions.

Throughout his career, Kelly has demonstrated an exceptional ability to connect with individuals from all walks of life, fostering a culture of collaboration and trust that empowers teams to excel. This biography takes you on an odyssey through Kelly's life, highlighting the key milestones that have defined his path to leadership.

You'll witness the challenges he faced, including the high-stakes environment of the aerospace industry, the fallout from the 737 Max crisis, and the labor relations hurdles that have tested his resolve.

Yet, amid these trials, Kelly's steadfast belief in the power of open communication and empathy has emerged as a beacon of hope, guiding his decisions and inspiring those around him. As you delve into the pages that follow, you'll discover not just the professional achievements of Kelly Ortberg but also the personal philosophy that shapes his leadership style.

You'll gain insight into the values that drive him—integrity, social responsibility, and a deep commitment to continuous learning. This biography serves as a testament to how one individual can make a lasting impact on an industry and the lives of countless others.

Join us as we explore the life of Kelly Ortberg, a leader whose vision and dedication are transforming Boeing and inspiring the next generation of innovators in aerospace and beyond. Prepare to be captivated by the story of a man who embodies resilience, vision, and the relentless pursuit of excellence.

Chapter One; Early Life and Education

Kelly Ortberg's journey to becoming the CEO of Boeing didn't begin in boardrooms or executive suites. It started in the small Midwestern town where he was born and raised, shaped by a tight-knit community and the values of hard work, integrity, and perseverance.

Growing up in Cedar Rapids, Iowa, Ortberg was surrounded by the kind of people who made things with their hands—farmers, factory workers, and craftsmen. From an early age, he developed a curiosity about how things worked, a fascination that would drive him toward a career in engineering and leadership.

His childhood was a blend of typical American experiences—playing in the fields, doing chores, and watching his father work long hours to provide for the family. His father, a mechanic, had a profound influence on him.

Kelly would often sit in the garage, watching his dad tinker with engines, learning how to fix things, and understanding the value of persistence. These experiences planted the seeds of problem-solving and innovation in his young mind, qualities that would later become key components of his professional ethos. In school, Ortberg was a diligent student but not the kind of person to boast about it.

His friends and teachers remember him as someone who would help others quietly, without seeking the spotlight. He excelled in math and science, subjects that captured his imagination and provided a glimpse into the world beyond his small-town life.

However, what truly made him stand out wasn't just his academic aptitude—it was his ability to stay grounded, always remaining approachable and humble, even as he gained more recognition. After high school, Kelly attended the University of Iowa, where he majored in mechanical engineering. College, for him, wasn't just about lectures and labs; it was a time of self-discovery.

He wasn't simply focused on getting good grades but on understanding how his education could be applied to the real world. His passion for aerospace engineering took root during these years, sparked by a summer internship at Rockwell Collins, a company known for its contributions to aviation technology.

That internship changed everything. Kelly had always been fascinated by planes—how they worked, how they stayed in the air—but it wasn't until he worked with seasoned engineers at Rockwell Collins that he realized this was more than just a passing interest. He saw the future of aviation, not just as an industry but as a field that could push the boundaries of human capability.

The experience motivated him to focus his studies on aerospace and defense technology, laying the foundation for the career that would eventually take him to the highest ranks of Boeing.

Ortberg's time at the University of Iowa wasn't all smooth sailing. Like many students, he struggled at times to balance the heavy academic load with personal challenges. But rather than becoming discouraged, he leaned into the challenges. His resilience during those tough moments would later become a defining trait of his leadership style.

He learned how to manage pressure, how to make decisions when the stakes were high, and, most importantly, how to stay focused on the end goal even when things seemed overwhelming.

After graduating with a degree in mechanical engineering, Ortberg began working full-time at Rockwell Collins, the same company where his career had been sparked during his internship. He started in a junior engineering role, where he worked on designing cockpit systems for military aircraft. It was here that his technical expertise began to merge with his natural leadership abilities.

He wasn't content to just be an engineer; he wanted to understand how the entire operation worked—from product development to customer relations. Ortberg's education didn't stop with his formal degree.

As he advanced in his career, he continued to learn, not just through books or courses but from the people around him. He was always observing, always asking questions. He learned from seasoned engineers, from executives, and from the factory workers on the shop floor. This habit of lifelong learning would serve him well in the years to come, particularly when he would later face some of Boeing's most challenging moments.

Though his career eventually took him to some of the most high-stakes boardrooms in the aerospace industry, Ortberg never forgot the lessons he learned during those early years.

The discipline he developed in his studies, the hands-on experience he gained in engineering, and the humble beginnings that shaped his values—all of these formed the foundation of a leader who would eventually guide Boeing through some of its most turbulent times. Kelly Ortberg's early life and education weren't just about academic achievements; they were about the formation of character. It was during these years that he learned the value of hard work, the importance of humility, and the power of persistence.

Those values stayed with him, helping him navigate not just the technical challenges of aerospace engineering but the human challenges of leading one of the world's largest companies through times of crisis.

As he often says, it's not just what you learn in school that matters, but how you apply those lessons to life, work, and leadership. That mindset, cultivated during his formative years, would later be the key to his success at Boeing.

Chapter Two; Path to Leadership

Kelly Ortberg's rise to leadership didn't happen overnight. It was a journey marked by dedication, risk-taking, and a deep understanding of both the technical and human aspects of business.

After starting his career as an engineer at Rockwell Collins, Ortberg could have easily settled into the comfortable rhythm of a steady career in aerospace. But that wasn't his style. From the beginning, Kelly had a hunger to make a bigger impact, not just on the products his company developed, but on the way it operated and evolved in a rapidly changing industry.

The early years at Rockwell Collins were formative. Kelly was thrust into projects that tested not only his engineering skills but also his ability to think holistically about how things worked.

It wasn't enough to design a product—he had to understand how it fit into the larger system, how it met the needs of the pilots using it, and how it aligned with the company's long-term goals. This attention to detail and the ability to zoom out and see the bigger picture quickly set him apart from his peers. But it wasn't just his technical acumen that marked him for leadership.

Kelly had an innate ability to connect with people, whether it was the workers on the factory floor or the executives in the boardroom. He understood that the best leaders were those who knew how to listen.

Early on, he developed the habit of walking through the production lines, talking to workers, asking them questions about their processes, and genuinely seeking their input. He didn't do this out of a sense of obligation, but because he believed that the people closest to the work often had the best ideas for how to improve it. That ability to listen, combined with his growing technical expertise, began to open doors for Kelly.

His supervisors noticed his talent not just for engineering, but for leading teams and managing complex projects. He was promoted to various managerial positions where he was responsible for overseeing critical aspects of the company's product development.

As he moved through these roles, Kelly proved time and again that he could balance the technical demands of the job with the leadership challenges of managing people, budgets, and timelines. One of the turning points in Kelly's career came when he was tasked with leading the development of new avionics systems for military aircraft. It was a high-pressure role that demanded precision, innovation, and coordination between multiple departments.

The stakes were high, and failure was not an option. Kelly thrived in this environment. He brought a sense of calm to his team, even when the pressure mounted. He knew how to break down complex problems into manageable pieces, and his team respected him for it.

They knew they could count on him to steer the ship, no matter how rough the waters. As Kelly continued to rise through the ranks at Rockwell Collins, it became clear that his ambitions went beyond technical leadership. He began to take on more strategic roles within the company, eventually being named Executive Vice President and Chief Operating Officer.

In this role, Kelly was responsible for overseeing all aspects of the company's operations, from product development to customer relations.

It was a significant leap from his engineering roots, but Kelly approached it with the same curiosity and determination that had defined his entire career. He didn't just sit in his office reviewing reports or attending meetings. He made it a point to stay connected to the day-to-day operations of the company. He would often spend time with employees at various levels, from engineers to sales teams, understanding their challenges and celebrating their successes.

Kelly believed that the best leaders were those who never lost touch with the people doing the work, and this philosophy served him well as he took on more responsibility.

One of Kelly's defining moments as a leader came during the acquisition of Rockwell Collins by United Technologies. It was a massive, $30 billion deal that would reshape the aerospace industry. Kelly played a key role in guiding Rockwell Collins through the transition, ensuring that the company's culture and values remained intact even as it became part of a larger conglomerate.

This was no easy feat. Mergers of this scale are often fraught with tension, as employees worry about job security and cultural shifts. But Kelly managed the transition with grace, ensuring that his team felt supported and valued throughout the process.

After the acquisition, Kelly's leadership abilities didn't go unnoticed. When Boeing came calling, it was clear that Kelly was ready for the challenge. His path to leadership at Boeing was a natural extension of everything he had learned and accomplished at Rockwell Collins. He was more than just a skilled engineer or a competent manager—he was a leader who understood the importance of trust, communication, and vision.

At Boeing, Kelly faced a different set of challenges. The company was reeling from a series of crises, and it needed a steady hand to guide it back to stability. Kelly didn't shy away from the challenge.

He approached his role at Boeing with the same focus and determination that had defined his career, always mindful of the fact that leadership is not just about making decisions but about bringing people along with you on the journey. Kelly Ortberg's path to leadership was not a straight line. It was a series of challenges, opportunities, and lessons learned along the way.

But through it all, one thing remained constant: his belief that leadership is about serving others, about listening as much as directing, and about never losing sight of the human element in any organization.

Chapter Three; The Rise to CEO of Boeing

Kelly Ortberg's ascension to the role of CEO at Boeing was not simply the result of corporate ambition or careful career planning.

It was the culmination of years of dedication, overcoming obstacles, and a leadership style deeply rooted in authenticity. At every step of his journey, Ortberg built his reputation not by seeking the spotlight, but by consistently focusing on solving problems and lifting others around him. By the time Kelly Ortberg arrived at Boeing, he was already a seasoned executive.

His years at Rockwell Collins, where he worked his way from an engineer to CEO, had prepared him for the immense responsibility of leading a company with such global reach.

But stepping into Boeing during a period of turbulence would test everything he had learned up to that point. Ortberg joined Boeing during one of its most difficult chapters. The company had faced multiple crises, from the high-profile problems with the 737 Max to mounting financial woes. It was a time when Boeing's reputation, both in the industry and among the public, had taken a serious hit.

It wasn't just a company struggling with technical failures—it was a company at risk of losing the trust of its customers, employees, and stakeholders.

When Kelly Ortberg was named CEO, it wasn't just about filling a role—it was about steering a massive ship through a storm and finding a way to rebuild the confidence that had been lost. In his early days as CEO, Ortberg was remarkably hands-on. Rather than staying cloistered in the executive suite, he took time to visit Boeing's factories, talking to workers on the shop floor, engineers in their offices, and even taking the time to meet with the families of employees who had been impacted by Boeing's struggles.

His approach wasn't about issuing directives from a distance—it was about understanding the pain points, fears, and frustrations of those who made Boeing tick.

He wasn't interested in maintaining a facade of corporate strength; he wanted to get to the core of what was really happening on the ground. One of the first significant challenges Ortberg faced was restoring Boeing's credibility with its customers and regulators. The 737 Max disaster had shaken the industry. The grounding of the aircraft after fatal crashes raised deep concerns about safety, engineering, and oversight at Boeing.

For Ortberg, this wasn't just about fixing a technical problem—it was about restoring faith in Boeing's ability to produce safe, reliable aircraft. Kelly approached this issue with a sense of urgency, but also with deep care.

He made transparency a top priority, committing to open communication with regulators, airlines, and even the families affected by the tragedies. This was not an easy path. In many ways, Boeing was under intense scrutiny, and every decision Kelly made was closely watched. But he understood that for Boeing to truly recover, it needed to own its mistakes and demonstrate a clear, actionable plan for moving forward.

Internally, Ortberg knew that Boeing needed more than just technical solutions. The company had to heal from within. Years of cost-cutting measures, workforce reductions, and intense corporate pressure had eroded the morale of Boeing's employees.

For Kelly, leading Boeing wasn't just about fixing the products—it was about rebuilding a culture that had suffered in the wake of these crises. He focused on fostering open communication within the company, encouraging feedback from employees at all levels. He wasn't afraid of tough conversations. In fact, he welcomed them, knowing that rebuilding Boeing's internal trust was just as important as regaining the confidence of the public.

He initiated programs to strengthen Boeing's culture of safety, innovation, and accountability, ensuring that every employee felt a part of the company's mission to restore its standing.

During his tenure, Ortberg also had to grapple with Boeing's financial challenges. The company was heavily in debt, and it hadn't turned an annual profit in years. Pressure to cut costs and streamline operations was immense, but Kelly knew that short-term cuts wouldn't solve the long-term issues. He was strategic about where Boeing needed to make sacrifices and where it needed to invest, particularly in talent and innovation.

Ortberg had a vision not just to get Boeing out of the hole, but to make it a leader again in a highly competitive aerospace industry. One of Kelly's strengths as a leader was his ability to remain calm under pressure.

He often described his leadership style as that of a steady hand, guiding the company through turbulent times without losing sight of the bigger picture. Even when faced with difficult decisions, such as laying off workers or pausing production on key projects, Ortberg approached every challenge with thoughtfulness, considering the human impact of every move.

Kelly Ortberg's rise to CEO of Boeing was a testament to his resilience, adaptability, and deep understanding of both the business and the people who work within it.

While his journey to the top wasn't free from challenges, he remained grounded in the principles that had defined his career—integrity, empathy, and an unwavering commitment to excellence. He understood that Boeing's recovery would take time, but he never shied away from the hard work required to restore its reputation and rebuild its future.

By the time Kelly stepped into the role of CEO, he had already proven that he wasn't just a leader for the good times—he was a leader for the stormy seas.

And at Boeing, those qualities would be tested to their fullest, shaping the legacy of a man who never lost sight of what truly mattered: the people behind the planes, and the trust that binds a company to its customers.

Chapter Four; Challenges of a New CEO

When Kelly Ortberg took the reins as CEO of Boeing, he stepped into a role laden with enormous expectations, but also fraught with challenges that could shake even the most seasoned leader.

Transitioning into the position of CEO isn't simply about learning the mechanics of the business. It's about stepping into the heart of a company's identity, tackling internal and external crises, and managing the weight of a legacy that has taken decades to build, but can be dismantled in mere moments. For Kelly, the challenges were immediate.

Boeing was in the midst of one of its most tumultuous periods, recovering from the 737 Max disasters that not only resulted in tragic loss of life but severely damaged the company's reputation for safety and reliability.

With the 737 Max fleet grounded globally, the financial and reputational fallout hung heavily over the company. Customers, regulators, and even the flying public were looking to Boeing for answers, and Kelly was at the forefront of this scrutiny. One of the first challenges for any new CEO is establishing credibility, and for Ortberg, this was critical. Boeing's name had been dragged through the mud, and trust, once shattered, is not easily rebuilt.

He faced a barrage of questions: How would he fix the problems with the 737 Max? What would he do to prevent something like this from ever happening again? And most importantly, could he restore Boeing's good standing in the eyes of its stakeholders? Kelly's approach to these challenges was deeply human.

Instead of sweeping the problems under the rug, he confronted them head-on. He began by fostering a culture of transparency, not just internally but with regulators, customers, and the public. He opened the doors for third-party scrutiny, inviting investigators and aviation authorities to review Boeing's processes and systems. This level of openness was a stark contrast to the closed-off approach the company had been accused of in the past.

For Ortberg, healing the wounds meant taking full responsibility and letting Boeing's actions, not just its words, speak for themselves. Another challenge Kelly faced as the new CEO was mending the internal fractures within Boeing.

Years of corporate missteps and financial pressures had created an environment where employees feel undervalued and disconnected from the company's mission. The pressure to perform in the face of tight deadlines and shrinking budgets had strained relationships within teams, from the engineers on the ground to the executives in the boardroom. Kelly understood that Boeing couldn't simply build better planes without also rebuilding its internal culture.

One of his first moves was to meet with employees across various levels of the organization. He listened to their concerns, frustrations, and suggestions, acknowledging that the people closest to the work often had the clearest insight into what was going wrong.

This wasn't just a box-ticking exercise for him. Kelly believed that the key to turning Boeing around lay in empowering its people again—helping them to feel that they were part of the solution, not just cogs in a corporate machine. Financial pressures were another significant hurdle. Boeing's financial health was teetering, and as the new CEO, Ortberg was expected to pull the company out of its dire straits.

The 737 Max crisis had drained billions from Boeing's reserves, and the company was saddled with massive debt. Kelly was tasked with stabilizing Boeing's finances while simultaneously addressing the company's operational and safety challenges.

Balancing these priorities was no small feat. Ortberg initiated a series of cost-cutting measures, including temporary furloughs, layoffs, and streamlining operations to reduce waste. But he also knew that cutting costs alone wouldn't be enough to secure Boeing's future. He began to explore new avenues for growth, particularly in the military and space sectors, where Boeing still had strong relationships and untapped potential.

Diversifying the company's revenue streams became a key part of Kelly's strategy, ensuring that Boeing wasn't putting all its eggs in one basket as it worked to recover from the 737 Max fallout. Externally, the pressure from regulators was immense.

The FAA, along with aviation authorities worldwide, had lost confidence in Boeing's ability to ensure the safety of its aircraft. Kelly had to navigate the delicate process of rebuilding trust with these entities. This meant ensuring that Boeing's new aircraft designs would meet the highest safety standards, while also dealing with the logistical nightmare of getting the 737 Max fleet back in the air.

Every delay cost the company millions, but Kelly knew that rushing the process would be far worse in the long run. In addition to dealing with financial and regulatory pressures, Kelly faced the challenge of steering Boeing through a rapidly changing industry.

The aerospace sector was evolving, with new players like SpaceX and Blue Origin pushing the boundaries of what was possible. Boeing, a historic titan in the industry, couldn't afford to rest on its laurels. It had to innovate, and Kelly had to inspire his team to embrace a future that looked very different from the company's past. For any new CEO, the first year is often about learning the ropes, but for Kelly Ortberg, it was about survival.

He had to make decisions that would not only define his legacy but also determine the future of Boeing. The stakes were incredibly high, and the challenges he faced were monumental, but Kelly remained steadfast.

He knew that leadership in such times wasn't about quick fixes, but about rebuilding trust, reinvigorating teams, and setting a clear path forward. Kelly Ortberg's first year as Boeing's CEO was marked by crisis management, but also by hope—hope that through transparency, accountability, and human-centered leadership, Boeing could rise once again.

Chapter Five; The Boeing 737 Max Crisis

The Boeing 737 Max crisis is a deeply human story of tragedy, corporate failure, and the long road to redemption.

When Kelly Ortberg stepped into the role of CEO, he inherited a crisis that shook Boeing to its very foundation—a crisis that had cost lives, shattered trust, and left the company's future in jeopardy. What was once seen as the future of aviation, the 737 Max, became synonymous with one of the worst periods in Boeing's history.

The story of the 737 Max began with promise. It was designed to be more fuel-efficient, with cutting-edge technology meant to keep Boeing at the forefront of commercial aviation. But in October 2018, just months after the 737 Max took to the skies, tragedy struck.

Lion Air Flight 610, a 737 Max, crashed into the Java Sea shortly after takeoff, killing all 189 passengers and crew. The world mourned, and Boeing began what should have been an urgent investigation into the cause of the crash. But before those answers could come, another disaster occurred. In March 2019, Ethiopian Airlines Flight 302, also a 737 Max, crashed, killing all 157 people on board.

The cause of both crashes was traced to a system called MCAS (Maneuvering Characteristics Augmentation System), designed to prevent the plane from stalling by adjusting the nose of the aircraft during certain flight conditions.

Tragically, in both crashes, the MCAS system malfunctioned, forcing the planes into a fatal dive that the pilots could not override. The world was in shock. How could a company with Boeing's reputation for safety produce a plane that allowed such a system to fail not once, but twice, with catastrophic consequences? Governments around the globe grounded the entire 737 Max fleet, and Boeing found itself at the center of intense scrutiny.

For the families of the victims, the crashes were a devastating loss. For Boeing, it was the beginning of a long, painful reckoning. As the newly appointed CEO, Kelly Ortberg's challenge was monumental.

This wasn't just about fixing a technical problem; it was about restoring faith in a company that had been the gold standard for safety and innovation. The stakes couldn't have been higher. Airlines canceled orders, customers lost trust, and regulators questioned Boeing's integrity. Ortberg knew that Boeing had to face this crisis with transparency and accountability. The first step was to admit where the company had gone wrong.

Internal investigations revealed a series of troubling failures in Boeing's development and certification of the 737 Max. The MCAS system had been rushed through testing, and Boeing had downplayed its importance to airlines and pilots, leaving many in the dark about how to respond to the system's failure.

It was a sobering discovery that led to intense soul-searching within the company. One of the most difficult challenges Kelly faced was restoring Boeing's internal culture. Years of cost-cutting and pressure to meet deadlines had led to an environment where safety concerns were not given the attention they deserved. Engineers had voiced worries about the MCAS system, but those concerns were either dismissed or not escalated.

Ortberg had to rebuild a culture where safety was paramount, not just a talking point in corporate speeches. He met with employees at all levels, urging them to speak up if they saw something wrong.

He wanted to restore a sense of pride in Boeing's mission: to make flying safe for everyone. The financial fallout from the crisis was immense. Boeing had to halt production of the 737 Max, which had been its best-selling plane. The company faced billions of dollars in compensation claims from airlines and suppliers, as well as lawsuits from the families of the victims.

The costs kept mounting, and with each passing day that the planes remained grounded, Boeing's reputation took another hit. Ortberg made difficult decisions to cut costs, furlough workers, and streamline operations, but he knew that no amount of financial restructuring could fix what had truly been broken—trust.

The process of getting the 737 Max back in the air was long and grueling. Regulators from around the world scrutinized every aspect of the plane's design, and Boeing worked tirelessly to fix the flaws in the MCAS system. Ortberg made it clear that this time, Boeing would do everything by the book, with no shortcuts. Every test, every safety measure, every protocol would be followed to the letter.

It wasn't just about getting the planes back in the air—it was about showing the world that Boeing had learned from its mistakes and was committed to earning back its place as a leader in aviation safety.

After months of rigorous testing and regulatory reviews, the 737 Max was finally cleared to fly again. For Ortberg and Boeing, it was a bittersweet moment. The planes were back, but the scars of the crisis would never fully heal. Families who had lost loved ones in the crashes were still grieving, and Boeing's name had been forever tarnished by the tragedy. Kelly Ortberg handled the crisis with empathy and determination.

He understood that Boeing's path to redemption would not be easy, but he remained focused on what mattered most: ensuring that something like this would never happen again.

The 737 Max crisis was a defining moment in Boeing's history, and under Ortberg's leadership, the company began the long journey of rebuilding both its reputation and its soul. For the world, the 737 Max crisis serves as a stark reminder that in the pursuit of progress, safety must always come first. And for Boeing, it was a humbling lesson that no matter how big or powerful a company may become, it is ultimately responsible for the lives of those who trust its products to carry them safely across the skies.

Chapter Six; Restoring Boeing's Reputation

Restoring Boeing's reputation after the 737 Max crisis was a monumental task that required more than just technical fixes—it demanded a change in mindset, culture, and, most importantly, trust.

For Kelly Ortberg, the newly appointed CEO, this wasn't just about leading a corporate turnaround; it was about healing the wounds left by a tragedy that had affected families, employees, and stakeholders alike. Restoring Boeing's once-stellar reputation was a journey of humility, accountability, and perseverance.

When Kelly stepped into the leadership role, Boeing was a company reeling from the twin crashes of the 737 Max, which had claimed the lives of 346 people.

It wasn't just a technical failure; it was a humanitarian disaster that had shattered the public's confidence in one of the most trusted names in aviation. As Boeing faced lawsuits, regulatory scrutiny, and massive financial losses, Kelly knew that simply getting the planes back in the air wasn't enough. Boeing had to regain the trust of passengers, airlines, and the global aviation community—something that had been painstakingly built over decades and lost in a matter of months.

The first step in restoring Boeing's reputation was admitting that things had gone terribly wrong. For a company of Boeing's stature, this was no easy task. The pressure to remain competitive had led to missteps, and it became clear that Boeing had prioritized speed and cost-efficiency over safety.

Under Kelly's leadership, Boeing made a public commitment to transparency. This meant opening up about the internal failings that had contributed to the crashes and, most importantly, taking responsibility for the mistakes that had been made. It was a difficult but necessary first step. Kelly understood that rebuilding Boeing's reputation had to start with its employees. The crisis had exposed a deeper problem within Boeing's corporate culture.

Over the years, cost-cutting measures and pressures to meet production timelines had created an environment where safety concerns weren't always prioritized, and where some employees felt hesitant to speak up.

To address this, Kelly made it clear that safety and transparency would no longer be negotiable. He introduced a series of changes aimed at empowering employees, encouraging them to voice concerns without fear of retribution, and reinforcing the company's core values. Boeing's engineers, designers, and factory workers were the heartbeat of the company, and they needed to be at the forefront of this transformation.

Internally, Kelly fostered open dialogue, holding town halls and creating new feedback mechanisms. Employees were encouraged to report safety issues, and the leadership team took the time to listen.

These efforts were about more than just making sure the planes were safe—it was about shifting the culture so that safety became the company's driving force once again. For employees, it wasn't just a top-down directive; it was a call to take pride in their work, knowing that the lives of millions of passengers depended on their vigilance and expertise. Externally, the path to restoring trust was even more challenging.

Airlines, regulators, and the flying public were understandably cautious about returning to Boeing's planes. Kelly and his leadership team worked tirelessly with regulators around the world, ensuring that the recertification process for the 737 Max was thorough and transparent.

Boeing didn't just want to meet safety standards—they wanted to exceed them. The company worked with the Federal Aviation Administration (FAA) and other global aviation bodies to review every aspect of the 737 Max's design and testing, implementing changes to the MCAS system and other key components. This time, there would be no shortcuts. Restoring relationships with airlines was another critical piece of the puzzle.

Many of Boeing's key customers had been deeply impacted by the grounding of the 737 Max, both financially and operationally. Kelly personally met with airline executives, not only to apologize for the disruptions but also to outline Boeing's plan for moving forward.

He listened to their concerns, responded to their questions, and made it clear that Boeing was committed to regaining their trust. Airlines needed assurance that they could rely on Boeing once again—not just in terms of safety but also in terms of reliability and support. Perhaps the most emotional and challenging aspect of restoring Boeing's reputation was addressing the families of the crash victims. For Kelly, this wasn't a business issue—it was a deeply human one.

He knew that no amount of apologies or compensation could ever bring back the loved ones lost in the crashes. But Boeing had to take responsibility and be part of the healing process.

Kelly ensured that Boeing reached out to the families, offering support and working closely with legal teams to resolve lawsuits with dignity and respect. The company also pledged to support aviation safety initiatives in honor of the victims, ensuring that their legacy would be one of positive change. Financially, Boeing had to rebuild confidence in the market as well. The company had suffered immense losses due to the grounding of the 737 Max, and shareholders were understandably anxious.

Under Kelly's leadership, Boeing implemented cost-saving measures, restructured its operations, and focused on long-term recovery. It wasn't an easy road, but Kelly remained steadfast in his commitment to restoring Boeing's financial health without sacrificing the renewed emphasis on safety and quality.

As the 737 Max was cleared to fly again and airlines began placing new orders, there was a palpable sense that Boeing was on the path to redemption. The crisis had left deep scars, but it had also prompted profound reflection and change. Kelly Ortberg's leadership throughout this challenging period demonstrated that restoring a company's reputation isn't just about fixing what's broken—it's about rebuilding the trust that allows people to believe in your brand again.

For Boeing, the 737 Max crisis became a lesson in humility, and under Kelly's guidance, the company embraced the opportunity to emerge stronger, safer, and more transparent than ever before.

Chapter Seven; Financial Struggles and Debt Crisis

When Kelly Ortberg assumed leadership as CEO of Boeing, the company was already grappling with an immense challenge: financial instability.

But what quickly became clear to Ortberg was that Boeing's financial troubles were far deeper than anyone had anticipated. The 737 Max crisis had drained the company's resources, and the global pandemic had delivered a crushing blow to the entire aviation industry. These challenges set the stage for a financial struggle that would test Boeing's resilience, leadership, and ability to recover.

In the years leading up to the crisis, Boeing was a dominant player in the aerospace industry, a titan known for its innovation and market presence. However, behind the scenes, the company was taking on massive debt to fuel its growth.

By the time Ortberg took over, Boeing's debt had swelled to a staggering $60 billion. This was the result of several factors: delays in aircraft production, compensation to airlines for grounded planes, lawsuits from the 737 Max crashes, and a halt in production caused by the pandemic. As airlines around the world slashed their fleets and reduced orders, Boeing's revenues plummeted. The company was in survival mode.

For Ortberg, the financial challenge was twofold: he had to stop the bleeding while also finding a way to stabilize and rebuild Boeing's financial foundation. The most pressing issue was the mounting debt.

Boeing, once a powerhouse that could easily secure loans, now faced the harsh reality of its situation. Credit ratings agencies were circling, threatening to downgrade Boeing's debt to junk status—a move that would skyrocket borrowing costs and severely limit the company's financial flexibility. Every day, Boeing was burning through cash, and the road to recovery seemed long and uncertain. To make matters worse, Boeing's stock had taken a nosedive. Investors, once bullish on the company's prospects, had lost faith.

The pressure was on Ortberg and his team to restore confidence, but they knew that achieving this would require difficult decisions. Boeing needed to cut costs—and fast.

This meant layoffs, furloughs, and restructuring. Thousands of workers were let go, and those who remained faced uncertain futures as the company navigated these troubled waters. It was an incredibly painful process for everyone involved, and Ortberg made it a priority to be as transparent as possible about the need for these measures. But cost-cutting alone wouldn't be enough to dig Boeing out of its financial hole.

The company had to rethink its business strategy. Part of the problem had been Boeing's overreliance on the 737 Max, which, before the crashes, had been its most successful aircraft.

With production halted and orders canceled, Boeing had to find other revenue streams to keep itself afloat. Ortberg worked closely with the leadership team to diversify Boeing's portfolio, focusing on its defense and space divisions, which had been relatively untouched by the commercial aviation downturn. One of the most critical moments in Boeing's financial recovery came when Ortberg and his team sought government assistance.

Early in the pandemic, Boeing had been offered a federal bailout, but the company had initially refused, wary of the conditions that might be attached. However, as the situation worsened, Boeing had little choice but to seek financial help.

Ortberg made a calculated decision to accept government-backed loans, ensuring that Boeing had the liquidity to survive the crisis. While this move brought some relief, it also underscored the gravity of Boeing's financial situation. The company, once seen as too big to fail, was now reliant on outside help to stay afloat. As Boeing battled its debt crisis, Ortberg's leadership was marked by a focus on long-term recovery rather than short-term gains.

He understood that the company couldn't simply patch up the holes and continue as it had before. Boeing had to fundamentally change the way it operated. This meant focusing on leaner, more efficient processes, reducing waste, and being more disciplined about where resources were allocated.

The pandemic had exposed vulnerabilities in Boeing's supply chain, and Ortberg was determined to address them. Yet, for all the financial challenges, there was a personal side to this story. Behind the balance sheets and debt restructuring were the people—employees, investors, customers, and suppliers—whose lives were directly impacted by Boeing's struggles. Ortberg took the time to speak with workers on the factory floor, acknowledging the sacrifices they were making to help the company recover.

He met with suppliers who had been hit hard by Boeing's delays, working to rebuild trust and ensure that they remained loyal partners. And with customers, Ortberg was upfront about the challenges Boeing faced, while also outlining a clear path to recovery.

The road ahead was daunting, but under Ortberg's leadership, Boeing began to slowly regain its footing. The company secured new contracts for defense projects, continued its work on space exploration, and gradually resumed production of its commercial aircraft. The debt load, while still significant, became more manageable as Boeing took steps to restructure its payments and renegotiate terms with creditors. It was a delicate balancing act, but one that Ortberg handled with a steady hand.

By the time Boeing began to emerge from its financial struggles, the company was forever changed. The debt crisis had been a wake-up call, forcing Boeing to confront the realities of its past mistakes and chart a more sustainable course for the future.

For Kelly Ortberg, it was a journey that tested his leadership in ways he could never have anticipated. But through it all, he remained committed to the people and the company he was entrusted to lead, knowing that restoring Boeing's financial health was just one part of the larger mission to restore its reputation and legacy.

Chapter Eight; The 2023 Machinist Strike

In the fall of 2023, Boeing found itself at the epicenter of a major labor movement as more than 30,000 machinists walked off the job in a historic strike that shook the company and the aerospace industry.

The strike was not just a demand for higher wages; it was a poignant reflection of workers' frustrations over long-standing issues that had festered during years of turbulence at Boeing. For Kelly Ortberg, the newly appointed CEO, this labor dispute posed a formidable challenge that would test his leadership and commitment to both the workforce and the company's future.

As the strike began, workers gathered on picket lines outside Boeing's factories in the Pacific Northwest, their resolve unmistakable. The atmosphere was charged with a mix of determination and anxiety.

For many machinists, the strike represented not only a fight for better pay and benefits but also a stand against what they perceived as a corporate culture that had increasingly undervalued their contributions. As they stood shoulder to shoulder, the sense of camaraderie was palpable. Many shared stories of their sacrifices—foregoing vacations, skipping meals, and working overtime—all while watching the cost of living in the Seattle area skyrocket.

Among the striking workers was Jake Meyer, a mechanic who had dedicated years of his life to building airplanes. He expressed the sentiments of many when he said, "We love our jobs, but we can't afford to live.

We're striking for our families and our futures." Jake's words encapsulated the frustrations of countless employees who felt trapped in a cycle of financial strain, particularly in a region where housing prices had surged by over 140% in just a decade. As the strike progressed, it became evident that the machinists were not merely fighting for a raise; they were demanding recognition of their worth and the essential role they played in Boeing's success.

The heart of the matter lay in the union negotiations. Workers sought a substantial wage increase, one that would keep pace with inflation and the soaring cost of living.

The International Association of Machinists and Aerospace Workers District 751 had initially proposed a 40% increase over four years, but Boeing's management countered with an offer of 25%. The gap between the two sides represented not only a difference in numbers but also a divergence in values—a reflection of how much the company valued its workforce.

Negotiators from both sides often expressed frustration over the lack of meaningful dialogue, highlighting the broader challenges facing labor relations in modern industries. In the midst of the strike, Ortberg faced mounting pressure to resolve the impasse.

He knew that a prolonged work stoppage would have serious financial implications for Boeing, already burdened with debt. The company was losing approximately $50 million a day due to halted production and disrupted supply chains. But Ortberg also understood that he needed to balance the company's financial health with a commitment to fair labor practices. He had to navigate a complex landscape where the voices of the workers needed to be heard, and their concerns addressed.

In an effort to bridge the divide, Ortberg initiated open forums where employees could voice their concerns directly to management. These sessions aimed to foster transparency and rebuild trust, which had been eroded in the years leading up to the strike.

Ortberg personally attended some of these meetings, listening to workers' stories and acknowledging their hardships. He recognized that many employees were contemplating side jobs just to make ends meet during the strike, and he wanted to show that he genuinely cared about their well-being. As negotiations continued, the mood on the picket lines fluctuated between hope and uncertainty.

Workers rallied together, sharing food, providing each other with support, and even organizing social events to keep morale high. Union leaders emphasized the importance of solidarity, encouraging members to remain steadfast in their demands.

Burn barrels provided warmth during chilly nights, and volunteers prepared meals for those on the picket lines, transforming the strike into a community effort. The striking workers weren't just fighting for themselves; they were also standing up for future generations. Many machinists articulated a desire to create better working conditions for those who would follow in their footsteps. "This isn't just about us," one worker said.

"It's about ensuring that the next generation of machinists have a seat at the table and aren't faced with the same struggles." As the days turned into weeks, the strike began to attract national attention.

Media coverage highlighted the plight of the machinists and their fight for fair wages, resonating with a broader audience that was increasingly sympathetic to labor movements across various sectors. The Biden administration weighed in, urging both sides to reach a resolution that would benefit workers and the company alike. The call for a fair agreement echoed throughout the country, emphasizing the need for a sustainable future for the aerospace industry. After weeks of negotiations, progress was finally made.

Both parties recognized the urgency of reaching an agreement before the strike could inflict further damage on the company and its employees.

Ortberg, alongside union leaders, worked diligently to broker a deal that would address the machinists' concerns while allowing Boeing to regain stability. Ultimately, an agreement was reached that included a more substantial wage increase than initially offered and improvements to benefits that had long been a sticking point. As the machinists returned to work, there was a palpable sense of victory in the air.

The strike had united them in a way that few other experiences could, and they returned to their jobs not just as employees, but as a cohesive community with a shared sense of purpose.

Ortberg's leadership during this tumultuous time was marked by his willingness to engage with workers and confront the difficult realities facing Boeing. The 2023 machinist strike was more than just a labor dispute; it was a transformative moment for Boeing. It served as a reminder of the essential role that workers play in the success of any organization and highlighted the ongoing challenges within labor relations in the modern economy.

It As Boeing moved forward, the lessons learned from this strike would shape the company's culture, leadership approach, and commitment to its workforce for years to come.

Chapter Nine; Labor Relations and Union Negotiations

Labor relations and union negotiations are crucial elements in the fabric of any organization, particularly in industries like aerospace, where the stakes are high and the workforce is skilled and dedicated.

At the heart of these relationships lies a complex interplay between the needs and aspirations of workers and the operational goals of the company. For Boeing, navigating this terrain has been both challenging and essential, particularly in recent years as the company faced significant labor disputes and financial pressures.

In an era marked by growing economic uncertainty and shifting labor dynamics, the relationship between Boeing and its machinists has come under intense scrutiny.

Workers, many of whom have devoted their careers to building the company's renowned aircraft, have increasingly voiced their frustrations over stagnant wages, rising costs of living, and concerns about job security. The 2023 machinist strike served as a powerful reminder of the importance of labor relations and the need for meaningful negotiations between management and employees.

When it comes to union negotiations, emotions often run high. For many machinists, the stakes are not just about pay increases or benefits; they are about recognition, respect, and the assurance that their hard work is valued.

These sentiments were palpable during the strike when workers took to the picket lines, united in their quest for fair treatment. As they waved signs and chanted slogans, the sense of solidarity among them was invigorating. Each worker had their own story, their own sacrifices, and their own dreams for the future, which they carried with them into the negotiations. At the center of these discussions were the union leaders, whose role is to advocate for the rights and needs of the workers they represent.

The International Association of Machinists and Aerospace Workers District 751 played a pivotal role during the negotiations, fighting for a contract that addressed the growing concerns of their members.

Union representatives listened to their members, gathering feedback and input to ensure that the proposals presented to Boeing reflected the collective voice of the workforce. This grassroots approach helped to empower workers, reminding them that they were not alone in their struggles. On the other side of the table, Boeing's leadership faced its own set of challenges.

Under CEO Kelly Ortberg, the company had to strike a balance between addressing the legitimate concerns of the machinists while also considering the financial implications of the negotiations.

With Boeing still recovering from the financial aftermath of the 737 Max crisis and the pandemic's toll on the aviation industry, the pressure was immense. Ortberg knew that a lengthy strike would not only further strain the company's finances but could also impact its reputation in an industry that is built on trust and reliability. As the negotiations progressed, it became clear that both sides needed to engage in open and honest dialogue.

Ortberg and his team began to host meetings where workers could voice their concerns directly, fostering an environment of transparency. These sessions allowed machinists to share their stories, express their frustrations, and articulate their aspirations.

For Ortberg, it was a crucial opportunity to connect with employees on a personal level and show that he understood the challenges they faced daily. The negotiations were far from easy, often marked by tense moments and conflicting interests. Workers were asking for significant wage increases to keep pace with inflation and the rising cost of living, while Boeing's management had to consider the company's financial health and long-term sustainability.

The disparity between the two sides' positions became a source of frustration, with many workers feeling unheard and undervalued. Union representatives made it clear that the machinists were willing to fight for what they believed was fair, and the pressure mounted on both sides to find common ground.

During this critical period, stories of individual machinists began to emerge, humanizing the negotiation process. There was Jake, a seasoned mechanic who had devoted over a decade to Boeing. He expressed his love for the work but lamented how difficult it had become to make ends meet in the Seattle area, where housing costs had skyrocketed. Then there was Maria, a single mother who worked tirelessly to provide for her children, feeling the weight of financial insecurity as she juggled work and family responsibilities.

Their experiences became emblematic of the broader struggles faced by the workforce, emphasizing the urgent need for a fair resolution. As negotiations continued, both sides recognized the importance of compromise.

Workers understood that the company needed to remain viable, and management began to see the necessity of investing in their workforce. After weeks of discussions, a tentative agreement was reached that included a more substantial wage increase than initially proposed, alongside enhancements to benefits and job security measures. The news was met with jubilation among the workers, who felt their voices had been heard and their sacrifices acknowledged.

In the aftermath of the agreement, it became evident that the resolution was more than just a contract; it was a turning point for labor relations at Boeing.

The strike had served as a catalyst for change, prompting both management and the workforce to reevaluate their relationships and commitments to one another. Ortberg's willingness to engage directly with employees and the union's efforts to advocate for workers' rights demonstrated that constructive dialogue could lead to positive outcomes.

The events surrounding the 2023 machinist strike highlighted the complexities of labor relations and the power of collective action. They underscored the importance of fostering a culture of respect, transparency, and collaboration within organizations.

As Boeing moved forward from this experience, the lessons learned would undoubtedly shape its approach to labor relations, helping to create a more resilient and united workforce for the future. Ultimately, the strength of any organization lies in its people. The stories of the machinists, their struggles, and their victories serve as a reminder that labor relations are about more than just contracts and negotiations; they are about people coming together to create a better workplace for all.

The resolution of the 2023 strike marked a new chapter in Boeing's history, one that would hopefully lead to stronger partnerships between management and labor, fostering a shared commitment to excellence and a brighter future for everyone involved.

Chapter Ten; Handling a Workforce in Rebellion

Handling a workforce in rebellion is one of the most daunting challenges that any leader can face, especially in industries where skilled labor is vital to success.

At Boeing, the ongoing labor disputes, particularly the 2023 machinist strike, serve as a striking example of how a company must navigate the turbulent waters of employee dissatisfaction and unrest. In the face of such challenges, leadership requires not only strategic acumen but also a deep understanding of the human element that drives employees' motivations and frustrations.

As tensions escalated among Boeing's machinists, it became evident that the workforce was not just pushing for better pay and benefits; they were demanding respect and recognition for their hard work and dedication.

The decision to strike was not taken lightly; it represented a culmination of years of grievances and a collective realization that their voices needed to be heard. For many machinists, this was about more than just wages—it was about standing up for their dignity and the future of their families. During times of unrest, leaders must first seek to understand the root causes of discontent. For Kelly Ortberg, Boeing's new CEO, this meant engaging with employees on a personal level.

He recognized that the workers were not just numbers on a spreadsheet; they were individuals with unique stories, challenges, and aspirations. Ortberg began attending union meetings, listening to the concerns of machinists, and acknowledging the pressures they faced in an increasingly expensive Seattle housing market.

By humanizing the relationship between management and labor, he aimed to rebuild trust and foster a sense of partnership. One striking aspect of the strike was the solidarity displayed among workers. From picket lines to union meetings, machinists rallied around one another, sharing their personal experiences and hardships. Stories like those of Jake, a veteran mechanic who had been with Boeing for over a decade, resonated deeply with his colleagues.

He articulated the struggles of making ends meet, particularly with rising living costs. Then there was Maria, a single mother who balanced work and family while navigating financial uncertainty.

Their stories served to underscore the urgency of the situation, and this emotional weight became a catalyst for change. Recognizing the importance of empathy in handling a workforce in rebellion, Ortberg made it a priority to keep communication lines open. He established regular forums where workers could voice their concerns, ensuring that their feedback was not just heard but acted upon. This move helped to shift the narrative from one of conflict to collaboration.

By inviting employees to share their perspectives, Ortberg demonstrated that he valued their input and was committed to finding solutions. However, addressing employee grievances is not simply about listening; it requires action.

When the machinists expressed their desire for a significant wage increase, Ortberg had to weigh the company's financial realities against the legitimate needs of the workforce. The challenge lay in finding a compromise that would meet both the expectations of the machinists and the financial constraints of Boeing. This delicate balance required creativity and a willingness to explore new ideas. Negotiations can be fraught with tension, especially when both sides feel the weight of their respective responsibilities.

For union leaders, the pressure to secure a favorable deal for their members is immense. They know that workers are counting on them to advocate for their rights, and the fear of not delivering can be overwhelming.

Ortberg understood this dynamic, and during negotiations, he worked to create a collaborative atmosphere where both sides could engage in constructive dialogue. The goal was to move beyond entrenched positions and find common ground. As the strike wore on, it became increasingly clear that the company needed to address the morale of its workforce. Workers on the picket lines were not only fighting for better wages but also for a sense of belonging and respect within the organization.

Ortberg recognized that disengagement could have long-lasting implications, not just for the present but for the future of Boeing. He initiated programs aimed at boosting employee morale, emphasizing the importance of recognizing their hard work and contributions.

Simple gestures like providing meals for picketers and offering support for families affected by the strike helped foster goodwill and demonstrate a commitment to the workforce. Eventually, after weeks of negotiations and a deepening sense of urgency, a tentative agreement was reached. The resolution was not merely a win for the workers; it represented a pivotal moment for Boeing as a whole. By taking the time to listen, empathize, and engage, Ortberg helped to transform a rebellious workforce into a collaborative partner in the company's future.

This experience underscored the power of human connection in the workplace. In the aftermath of the strike, it was evident that handling a workforce in rebellion requires more than just strategic maneuvers; it demands a willingness to acknowledge the humanity of every employee.

The stories of machinists like Jake and Maria became woven into the fabric of the company's culture, serving as reminders of the collective strength that can arise when workers feel valued and respected. The challenges faced during the strike taught Boeing valuable lessons about the importance of fostering open communication, empathy, and collaboration in labor relations. As the company moved forward, it recognized that the path to success is paved with the voices of its workforce.

In the end, the experience not only strengthened the bond between management and labor but also laid the groundwork for a more resilient and united Boeing, capable of facing the future together. Handling a workforce in rebellion is not merely about resolving conflicts; it's about building a culture that honors the dignity of every individual and recognizes the shared goals that unite us all.

Chapter Eleven; Economic Impact of the Strike

The economic impact of a strike can ripple far beyond the immediate effects on a company, affecting local communities, supply chains, and even entire industries.

The 2023 machinist strike at Boeing, which saw thousands of skilled workers walk off the job, serves as a vivid illustration of how labor disputes can shape economic landscapes, provoke urgent discussions about fairness in the workplace, and highlight the delicate balance between corporate profitability and employee welfare. As the strike unfolded, Boeing faced significant financial strain.

Estimates suggested that the company was losing around $50 million a day, a staggering figure that underscored the severity of the situation. This loss didn't just impact Boeing's bottom line; it reverberated throughout its extensive network of suppliers and partners.

Many of these companies depend on Boeing's production schedules and volume, meaning that a halt in aircraft manufacturing could lead to cascading effects in the aerospace supply chain. Parts suppliers, logistics companies, and even local businesses that support the workforce felt the impact. Workers at Boeing, many of whom had devoted years to their craft, were not only striving for better wages but also for dignity in their work and assurance about their futures.

For many of them, the economic implications were personal and immediate. With the high cost of living in the Seattle area, workers were grappling with the reality that a prolonged strike could mean financial hardship, especially as health benefits were set to expire soon.

This scenario was not just theoretical; workers like Jake, a Boeing mechanic, were making plans to take side jobs in landscaping and food delivery to make ends meet during the strike. His story reflected a broader anxiety shared by many: the fear of losing not only their livelihoods but also their sense of purpose and pride in their work. The impact of the strike extended to the communities surrounding Boeing's manufacturing plants.

In Renton and beyond, local businesses that relied on the foot traffic and spending of Boeing employees faced uncertainty. Cafés, restaurants, and retail shops saw a significant drop in customers as machinists took to the picket lines.

For many of these small businesses, the strike posed a double-edged sword: they understood the workers' plight but felt the pinch of lost revenue at a time when they were still recovering from the economic effects of the pandemic. Beyond the immediate financial losses, the strike sparked conversations about the broader implications of labor relations and economic inequality.

The machinists' demands for higher wages were not just about their circumstances; they echoed a nationwide dialogue about fair compensation in an era where many workers feel left behind.

As inflation surged and the cost of living skyrocketed, the strike underscored the frustrations of a workforce that had been waiting for a significant pay adjustment for years. The economic stakes were high, and the machinists were determined to advocate for a more equitable share of the profits generated by their hard work. The strike also prompted a critical examination of Boeing's financial health.

With the company already grappling with a staggering $60 billion in debt and a history of financial instability, the labor dispute raised questions about how Boeing would navigate these challenges moving forward.

Investors and analysts began to scrutinize Boeing's ability to manage not only its production challenges but also the growing expectations of its workforce. A potential downgrade in the company's credit rating loomed if the strike dragged on, compounding the financial woes that already plagued the aerospace giant.

In the long term, the economic impact of the strike could have lasting effects on Boeing's labor relations and overall strategy. The lessons learned from this labor dispute could influence how the company approaches negotiations in the future.

Recognizing the growing labor movement across various industries, including the automotive and tech sectors, Boeing's leadership might need to adopt a more proactive stance toward addressing employee concerns and fostering a collaborative work environment. The response from the Biden administration also highlighted the strike's national significance.

Transportation Secretary Pete Buttigieg publicly expressed the hope that both sides would reach a resolution, emphasizing the need for a fair deal that not only benefited the workers but also positioned Boeing for recovery and growth.

This sentiment resonated with many Americans who understood that the health of the aerospace industry was tied to the economy at large. Ultimately, the economic impact of the 2023 machinist strike at Boeing serves as a powerful reminder of the interconnectedness of labor, industry, and community. It illustrates the human element behind economic decisions and the profound effects that labor disputes can have on the lives of individuals and the broader economy.

As the dust settled and negotiations reached a resolution, the stories of workers like Jake and Maria remained at the forefront of the narrative, reminding everyone that the fight for fair wages and working conditions is not just an economic issue but a matter of dignity and respect.

In reflecting on the strike's economic implications, it becomes clear that labor disputes are not isolated incidents; they are moments of reckoning that can shape the future of entire industries. The conversations ignited by the machinists' actions extended beyond the walls of Boeing, prompting discussions about labor rights, economic justice, and the responsibilities of corporations to their workers.

As the aerospace industry moves forward, it carries the lessons learned from this strike, with the hope of fostering a more equitable environment for all involved. The interplay between labor and economic health will continue to evolve, and the voices of workers will remain central to this ongoing dialogue.

Chapter Twelve; Furloughs and Cost-Cutting Measures

Furloughs and cost-cutting measures are often seen as necessary evils in the corporate world, especially during times of financial strain.

For Boeing, these strategies became a reality amid an economic landscape fraught with challenges, including rising debts and ongoing labor disputes. As CEO Kelly Ortberg navigated these tumultuous waters, he faced the daunting task of balancing the company's financial needs with the human element that drives its workforce. The decision to implement furloughs is never taken lightly.

It usually comes after extensive analysis and discussions, and at Boeing, the stakes were particularly high. With a workforce deeply embedded in the company's operations, a furlough impacts not just productivity but the very morale of employees who have dedicated their careers to building some of the world's most iconic aircraft.

Ortberg understood that announcing temporary layoffs would send shockwaves through the organization. It wasn't just a matter of financial strategy; it was about people—employees, their families, and the communities that rely on them. In the wake of the 2023 machinist strike, the decision to furlough tens of thousands of workers was driven by an urgent need to manage costs. With the strike disrupting production and creating uncertainty, Boeing faced a significant cash flow issue.

It was a precarious position, as the company had already accumulated around $60 billion in debt. Ortberg and his leadership team had to find ways to reduce expenses without jeopardizing the long-term future of the organization.

As Ortberg prepared to communicate these decisions, he was acutely aware of the emotional weight they carried. Employees who had dedicated years of their lives to Boeing were now faced with uncertainty about their futures. Furloughs mean more than just a temporary loss of income; they disrupt lives and create anxiety about job security. Many workers, like Jake, a longtime machinist, worried about how they would support their families.

The anxiety was palpable as employees discussed potential side jobs or even relying on savings to make it through. Ortberg reached out to the workforce through a series of town hall meetings, where he openly addressed the challenges the company faced.

He emphasized that these measures, though difficult, were intended to protect the long-term stability of Boeing. He shared stories of resilience and commitment from within the company, hoping to inspire hope amid uncertainty. While some employees appreciated his transparency, others felt anger and frustration. The reality was that many families depended on their Boeing salaries for their livelihoods, and the idea of furloughs felt like a personal attack on their dedication.

Cost-cutting measures often extend beyond furloughs. At Boeing, these included a hiring freeze, the suspension of non-essential projects, and a reevaluation of operational efficiencies.

The leadership team had to scrutinize every aspect of the business, from production lines to administrative expenses, seeking ways to tighten belts and improve financial health. Ortberg urged managers to consider innovative solutions that could minimize costs while maintaining quality and employee engagement. Despite the dire circumstances, some employees found creative ways to cope with the challenges posed by the furloughs and cost-cutting measures.

Community initiatives sprang up, as workers banded together to support one another. They organized food drives and shared resources, understanding that collective strength was vital during these uncertain times.

This sense of community reflected the deep-rooted camaraderie among Boeing employees, reminding everyone that they were in this together. As the furloughs took effect, Ortberg emphasized the importance of communication and support. He encouraged employees to reach out if they needed assistance, whether through financial counseling or job placement services. The company also aimed to provide updates about the situation, ensuring that workers felt connected to Boeing's ongoing efforts to resolve the crisis.

Regular check-ins and virtual forums allowed employees to voice their concerns and seek clarity about their futures, helping to alleviate some of the stress associated with furloughs. The broader economic implications of these measures were also significant.

As Boeing cut costs and furloughed workers, local businesses in the surrounding communities felt the impact. Cafés, shops, and service providers that relied on the spending of Boeing employees saw a decline in business, creating a ripple effect through the economy. The tight-knit nature of the region meant that many felt the strain of Boeing's challenges, emphasizing the interconnectedness of corporate decisions and local livelihoods. As the months went by, the hope was that these tough choices would lead to a more stable Boeing.

Ortberg understood that while immediate financial relief was crucial, the company's long-term health depended on maintaining employee morale and fostering a culture of resilience.

He worked diligently to ensure that, as furloughs ended and operations ramped back up, employees would return to an environment that valued their contributions and recognized the sacrifices made during the difficult times. Furloughs and cost-cutting measures, though often seen as a sign of distress, can also serve as an opportunity for reflection and renewal. For Boeing, this period was a chance to reassess its values and commitment to its workforce.

By prioritizing transparency, empathy, and community support, Ortberg sought to not only navigate the financial challenges but also to strengthen the bonds between management and employees.

Ultimately, the experience of handling furloughs and cost-cutting measures at Boeing illustrates the delicate balance leaders must strike between fiscal responsibility and human connection. It's a reminder that behind every decision, there are lives impacted and stories waiting to be told. As Boeing emerges from these challenges, it carries with it the lessons learned about the importance of compassion in leadership and the enduring strength of a united workforce.

The journey forward may be fraught with obstacles, but together, they can navigate the skies once more.

Chapter Thirteen; The Road to Resolution

The road to resolution is rarely a straight path; it's often winding, filled with challenges, setbacks, and moments of uncertainty.

This was especially true for Boeing during the tumultuous period marked by the 2023 machinist strike. The conflict between the company and its workers wasn't just about wages or benefits; it symbolized deeper issues related to employee trust, corporate accountability, and the future direction of one of the world's most iconic companies.

As the strike began, tensions ran high. Machinists, who had dedicated years to building Boeing's legacy, took to the picket lines fueled by frustrations over stagnant wages and rising living costs.

The strike was a culmination of pent-up grievances, a moment when workers felt their voices needed to be heard. For many, the picket line became a place of solidarity, a testament to their commitment not only to their craft but also to each other. The air was thick with determination, but as the days turned into weeks, uncertainty loomed like a shadow. Kelly Ortberg, the newly appointed CEO, found himself in the hot seat.

His leadership was under scrutiny not just from investors but also from the very workers who had once looked to Boeing as a beacon of stability. Ortberg was acutely aware that restoring the company's reputation and resolving the strike were intertwined.

It was essential to address the concerns of the workforce while also steering the company back to financial health. The dual pressures of external expectations and internal strife created a perfect storm that required skillful navigation. In the midst of the turmoil, Ortberg made it a priority to engage directly with the striking workers. He attended rallies and spoke with machinists, seeking to understand their perspectives firsthand. These interactions were pivotal.

They humanized the leadership for the employees, who often felt disconnected from decision-makers in the corporate offices. Ortberg listened to their stories, learned about their struggles, and acknowledged their sacrifices.

This approach was crucial in rebuilding trust, as many workers felt their voices had been ignored for far too long. The negotiations that followed were not easy. Both sides had to grapple with their respective demands and realities. The union represented thousands of workers, each with individual needs and concerns, while Boeing had to consider its financial constraints and the larger economic environment.

Ortberg emphasized transparency throughout the negotiation process. He regularly communicated updates to the workforce, sharing the challenges Boeing faced while outlining the company's commitment to finding a resolution.

As talks progressed, moments of tension were met with gestures of goodwill. Ortberg proposed mediation sessions, inviting third-party negotiators to help facilitate discussions. This move was met with cautious optimism. It indicated that Boeing was willing to compromise and find common ground, even amid disagreement. The presence of neutral mediators brought a fresh perspective, allowing both parties to explore creative solutions that might not have been considered otherwise.

Workers were eager for a resolution, not just to return to their jobs but to regain a sense of normalcy in their lives. The financial strain of the strike weighed heavily on many families, and discussions about side jobs and financial struggles permeated conversations among machinists.

Stories like Jake's, who had been delivering food to make ends meet, underscored the human cost of the conflict. Such narratives reminded everyone involved that behind every negotiation were real people facing real challenges. The breakthrough finally came when both parties reached a tentative agreement that addressed many of the key concerns. Boeing committed to substantial wage increases and improved benefits, and the union agreed to return to work while the details were finalized.

This moment of relief was palpable; it marked the end of a tumultuous chapter and the beginning of a new era. As workers returned to their jobs, the atmosphere shifted.

The factory floors buzzed with renewed energy, and there was a sense of camaraderie among the machinists. The resolution wasn't just about the terms of the agreement; it was about restoring faith in the company and reaffirming the bond between Boeing and its workforce. Ortberg recognized that the path to resolution required more than just contractual agreements; it necessitated ongoing dialogue and commitment to fostering a positive workplace culture.

In the months that followed, Ortberg focused on implementing the changes promised during negotiations. He established regular town hall meetings to encourage open communication, ensuring that employees felt valued and heard.

The company also invested in employee training and development programs, demonstrating a long-term commitment to its workforce. The road to resolution was not just a conclusion to a labor dispute; it represented a transformative moment for Boeing. The experience highlighted the importance of empathy in leadership and the need for companies to prioritize their employees' well-being.

Ortberg's approach to resolving the strike served as a blueprint for addressing future challenges, emphasizing the significance of trust and collaboration. In reflecting on the journey, it became clear that the road to resolution was not merely about returning to business as usual.

It was about rebuilding a company culture that valued every employee's contributions and aspirations. The lessons learned from the strike resonated throughout the organization, reminding everyone that behind the machinery and production lines were dedicated individuals whose passion and commitment fueled Boeing's success. As Boeing moved forward, the resolution marked a new beginning, one rooted in understanding and mutual respect.

The experience illustrated that, while the road to resolution may be fraught with obstacles, the collective strength of a workforce united by a common purpose can overcome even the most challenging of circumstances. The journey may have been tough, but the destination promised a brighter, more inclusive future for all at Boeing.

Chapter Fourteen; The Future of Boeing Under Ortberg

As Kelly Ortberg took the helm of Boeing amid one of the most tumultuous periods in the company's history, the future seemed uncertain. The challenges of the past—financial struggles, labor unrest, and the fallout from the 737 Max crisis—loomed large.

Yet, Ortberg's leadership style and vision for the future offered a glimmer of hope and a pathway forward for both the company and its employees. From the outset, Ortberg recognized that the road ahead would not be easy. He faced a workforce that had endured significant hardship, not just financially but also in terms of trust and morale. The machinists' strike had underscored deep-seated frustrations about wages and working conditions.

Ortberg understood that to move forward effectively, he needed to address these issues head-on. His commitment to open dialogue became a cornerstone of his approach.

By holding regular town hall meetings and encouraging feedback from employees, he aimed to create an environment where everyone felt valued and heard. Ortberg's vision for Boeing extended beyond immediate recovery; it was about laying the groundwork for sustainable growth and innovation. He believed that the key to revitalizing the company lay in its people. Investing in employee training and development was not just a strategic move; it was a heartfelt acknowledgment of the dedication and expertise that the workforce brought to the table.

Ortberg often shared stories of machinists who had spent decades honing their craft, emphasizing that their skills were the lifeblood of Boeing. This focus on people helped foster a sense of unity, reigniting the pride that many felt in their work.

In tandem with nurturing the workforce, Ortberg understood the necessity of innovation in a rapidly evolving aerospace landscape. The aviation industry was shifting, driven by technological advancements and changing customer demands. To ensure Boeing remained at the forefront, Ortberg championed a renewed emphasis on research and development. He advocated for partnerships with tech companies and universities, fostering a culture of collaboration that would propel Boeing into the future.

By embracing cutting-edge technologies like artificial intelligence and sustainable aviation fuels, Ortberg positioned Boeing not just as a manufacturer of airplanes, but as a leader in reshaping the future of flight.

Financial recovery was also paramount in Ortberg's strategy. With Boeing's debt levels soaring and the company grappling with significant losses, he prioritized restoring profitability. This required tough decisions, including cost-cutting measures and streamlining operations. However, Ortberg was careful to balance these necessary actions with a commitment to maintain a positive workplace culture. He understood that while financial stability was crucial, it could not come at the expense of the workforce's well-being.

Ortberg's leadership was further characterized by transparency. He regularly communicated Boeing's financial status and operational goals to employees, ensuring they understood the challenges ahead.

This openness not only fostered trust but also encouraged employees to take ownership of their roles in the company's recovery. It was clear that Ortberg wanted every machinist, engineer, and manager to feel like a vital part of the Boeing story, reinforcing the idea that their contributions mattered. As Boeing began to recover, Ortberg focused on rebuilding relationships with customers and stakeholders. Trust had been eroded during the crisis, and Ortberg knew that restoring faith in Boeing was essential for future success.

He made it a priority to engage directly with airline customers, seeking to understand their needs and concerns. By demonstrating a commitment to quality and safety, Ortberg aimed to reassure customers that Boeing was a reliable partner in their business.

The future of Boeing also required a renewed commitment to sustainability. As environmental concerns became increasingly prominent, Ortberg recognized the importance of aligning the company's goals with global sustainability initiatives. Boeing started investing in eco-friendly technologies and exploring ways to reduce emissions in its manufacturing processes.

Ortberg often spoke about the importance of being responsible stewards of the planet, emphasizing that sustainability was not just a trend but a necessity for the future of the aviation industry.

Under Ortberg's leadership, Boeing's trajectory began to shift. While the road was still challenging, there were signs of recovery and renewed energy within the company. Employees returned to their jobs with a sense of purpose, motivated by the knowledge that their work was integral to Boeing's mission. The atmosphere in the factories was charged with optimism, and the camaraderie among the workforce began to flourish once more.

Looking ahead, Ortberg's vision for Boeing was one of resilience and adaptability. He understood that the future would inevitably bring new challenges, but he believed that the company's rich history and skilled workforce positioned it well to navigate whatever lay ahead.

The foundation he was building—rooted in trust, innovation, and sustainability—promised not only to restore Boeing's reputation but to elevate it to new heights. In reflecting on Ortberg's leadership journey, it became clear that the future of Boeing was not solely about the products it created but about the people behind those products. The relationships forged between management and employees, the commitment to innovation, and the focus on sustainability all played vital roles in shaping a brighter future.

As the company moved forward, it embraced a vision that was inclusive, forward-thinking, and deeply connected to the values that had long defined Boeing as a leader in the aerospace industry.

Ultimately, the future under Kelly Ortberg was one marked by hope and possibility. As Boeing continued to navigate the complexities of the aviation landscape, it did so with a renewed sense of purpose—a commitment to excellence, a dedication to its people, and an unwavering belief in the transformative power of flight. The journey ahead would undoubtedly be challenging, but with Ortberg at the helm, Boeing was poised to soar once more.

Chapter Fifteen; Legacy in Aerospace Leadership

In the ever-evolving world of aerospace, legacy often becomes synonymous with leadership, shaping the industry through both vision and perseverance.

As the field has grown more complex, the leaders who have emerged at the forefront have not only crafted their companies' destinies but also redefined the very landscape of aviation. Their legacies are characterized by innovation, resilience, and a commitment to the people and technologies that propel the industry forward. One of the most prominent figures in aerospace history is Boeing's founder, William Boeing.

His journey began in the early 20th century when he ventured into aviation at a time when the concept of commercial flight was still in its infancy. Boeing's vision was simple yet profound: he believed in the transformative potential of flight.

His insistence on quality and innovation laid the groundwork for what would become a multi-billion dollar industry. Boeing's legacy is not just in the airplanes that bear his name, but in the spirit of exploration and advancement he instilled in his company and the industry at large. Fast forward to the late 20th century, and leaders like Frank Shrontz, former CEO of Boeing, emerged during a period of rapid change and uncertainty.

Under his stewardship, Boeing navigated the challenges posed by global competition and the complexities of mergers and acquisitions. Shrontz understood that collaboration and synergy were essential for survival in the aerospace sector.

His approach was rooted in the belief that a successful company thrives on the collective efforts of its workforce. By fostering a culture of teamwork and open communication, he helped Boeing adapt and flourish amid turbulent times, ensuring that the company remained a powerhouse in aerospace. Another key figure in aerospace leadership is the former Lockheed Martin CEO, Marillyn Hewson. Breaking barriers as one of the first women to lead a major defense contractor, Hewson's legacy is one of empowerment and diversity.

Under her leadership, Lockheed Martin not only expanded its market share but also committed to fostering an inclusive workplace. Hewson understood that innovation flourished in environments where diverse perspectives were encouraged.

She championed initiatives to recruit and retain female talent in STEM fields, thereby paving the way for the next generation of leaders in aerospace. Her legacy transcends corporate achievements; it reflects a fundamental shift in the industry's approach to inclusivity and representation. The legacies of these leaders also highlight the importance of ethical considerations in aerospace. In an industry often scrutinized for its environmental impact, leaders like Ray Conner, who served as CEO of Boeing Commercial Airplanes, have made strides toward sustainability.

Conner spearheaded efforts to develop more fuel-efficient aircraft and emphasized the need for corporate responsibility. His vision was clear: the future of aerospace must align with the principles of sustainability and environmental stewardship.

This forward-thinking approach not only addresses the industry's challenges but also sets a precedent for future leaders who will inherit the responsibility of balancing innovation with ecological considerations. Moreover, the legacies of aerospace leaders are intertwined with their ability to navigate crises. The impact of the 737 Max crisis on Boeing exemplifies this. Kelly Ortberg, stepping into the CEO role during tumultuous times, faced an uphill battle in restoring trust and reputation.

His approach to leadership emphasized transparency and open communication with employees, customers, and stakeholders. Ortberg's legacy is one of resilience, demonstrating that true leadership is revealed in moments of adversity.

His focus on rebuilding relationships and fostering a culture of accountability signifies a commitment to ethical leadership that resonates deeply within the industry. Legacy in aerospace leadership is also about embracing change and adapting to new technologies. Leaders like Elon Musk have transformed not only the perception of space travel but also the very nature of aerospace innovation. With SpaceX, Musk has made strides in reusability and cost reduction, revolutionizing the industry.

His vision of interplanetary exploration and the colonization of Mars pushes the boundaries of what is possible. Musk's legacy serves as a reminder that audacity and innovation can reshape industries and inspire future generations to dream bigger.

As the aerospace industry moves forward, the legacies of past leaders continue to inform and inspire. Emerging leaders must recognize the importance of ethical considerations, inclusivity, and innovation. They will inherit the challenge of navigating a rapidly changing landscape marked by technological advancements and environmental concerns. By learning from the successes and challenges of those who came before, they can forge paths that honor the legacies of pioneers while also shaping a brighter future.

In essence, the legacy of aerospace leadership is built on a foundation of vision, courage, and a commitment to the greater good. It is a narrative that intertwines the aspirations of individuals with the collective achievements of an industry.

The stories of leaders like William Boeing, Frank Shrontz, Marillyn Hewson, Kelly Ortberg, and Elon Musk remind us that leadership in aerospace is not merely about the products or profits; it is about people, relationships, and the impact of flight on humanity. As the world of aerospace continues to evolve, the legacies of these leaders will endure, inspiring a new generation to take up the mantle of leadership with a sense of responsibility and purpose.

The future of aviation is bright, fueled by the dreams and ambitions of those willing to push boundaries, embrace change, and make a lasting impact on the world. In this realm, every leader has the potential to leave an indelible mark, shaping not only their companies but also the future of flight itself.

Chapter Sixteen; Personal Life and Philosophy

Kelly Ortberg's personal life and philosophy are deeply intertwined with his professional journey, reflecting a man who understands the delicate balance between ambition and personal values.

Born and raised in a modest household, Kelly developed a strong work ethic early on, shaped by his parents' dedication to their jobs and their community. This foundation not only instilled a sense of responsibility but also fostered a belief in the importance of integrity and hard work. These values would later guide him through his career at Boeing and influence his leadership style. Kelly's upbringing in a small town taught him the significance of connection and community.

He often reflects on how his childhood experiences shaped his understanding of people and relationships. Family gatherings, school events, and community service projects were not just activities; they were opportunities to learn from others and to appreciate the diverse perspectives that people bring to the table.

This appreciation for community would become a hallmark of his leadership philosophy, emphasizing collaboration and mutual respect in the workplace. Education played a crucial role in Kelly's development. He pursued a degree in mechanical engineering, a choice that not only aligned with his interests but also equipped him with the technical skills necessary for a career in aerospace. During his time at university, he encountered challenges that tested his resolve.

Yet, he approached these challenges with a curious mind and a willingness to learn, often reminding himself that every obstacle is an opportunity for growth. This mindset laid the groundwork for his later achievements and decision-making processes.

As he climbed the ranks within Boeing, Kelly maintained a strong belief in the importance of lifelong learning. He often encourages his teams to embrace new ideas and perspectives, fostering an environment where innovation can thrive. His philosophy is simple yet powerful: learning never stops, and the pursuit of knowledge should be a continuous journey. This belief not only keeps him motivated but also inspires those around him to strive for excellence.

Outside of work, Kelly prioritizes family life, understanding that personal connections are just as important as professional accomplishments. He often shares stories of family vacations, holiday traditions, and weekend activities that allow him to unwind and recharge.

These moments are crucial for maintaining a healthy work-life balance, which he views as essential for overall well-being. He recognizes that being a successful leader is not solely about achieving business goals; it's also about being present for loved ones and nurturing those relationships. His leadership style is rooted in empathy and understanding. Kelly believes that a successful leader must be attuned to the needs and concerns of their team members.

He often takes the time to listen, ensuring that everyone feels valued and heard. This approach has fostered a culture of trust and collaboration within his teams.

By prioritizing open communication, he creates an environment where individuals feel comfortable sharing their ideas and concerns, ultimately leading to better decision-making and problem-solving. Kelly's personal philosophy also includes a commitment to social responsibility. He understands that as a leader in a global company, he has a responsibility to give back to the community and to support initiatives that make a positive impact.

Whether through corporate social responsibility programs or personal involvement in charitable organizations, he emphasizes the importance of making a difference.

His belief in the power of community service resonates throughout his leadership approach, encouraging employees to engage in volunteer work and contribute to causes they are passionate about. Resilience is another cornerstone of Kelly's philosophy. He firmly believes that setbacks are an inevitable part of any journey, and it is how one responds to those challenges that truly defines success. He often recounts his experiences during difficult times at Boeing, where he faced scrutiny and pressure.

Instead of shying away from challenges, he embraced them, using adversity as a catalyst for growth and improvement. This resilience has become a source of inspiration for many within the organization, demonstrating that challenges can lead to innovation and progress.

In a rapidly changing world, Kelly is committed to adaptability. He recognizes that the aerospace industry is evolving, and leaders must be willing to pivot and embrace new technologies and methodologies. His openness to change is not just about staying relevant; it's about fostering a culture of innovation that empowers others to think outside the box. By encouraging his teams to explore new ideas and approaches, he cultivates an atmosphere where creativity flourishes.

As Kelly Ortberg continues to navigate his leadership journey, his personal life and philosophy serve as a guiding compass. His commitment to integrity, lifelong learning, family, empathy, social responsibility, resilience, and adaptability shapes not only his career but also the culture at Boeing.

Through his actions and values, he inspires others to strive for excellence while remaining grounded in the principles that matter most. In the end, Kelly's story is a testament to the idea that true leadership goes beyond titles and accomplishments. It's about the impact one has on others, the values one embodies, and the legacy one leaves behind.

As he looks to the future, Kelly Ortberg remains dedicated to fostering a culture of innovation and inclusivity at Boeing, driven by the belief that success is best achieved through collaboration and a shared vision for a brighter tomorrow.

Conclusion

In closing, the journey of Kelly Ortberg is a remarkable testament to the power of resilience, innovation, and principled leadership in the face of adversity.

From his early life shaped by strong values to his rise through the ranks at Boeing, Kelly has exemplified what it means to lead with integrity and empathy. His commitment to fostering a culture of collaboration, open communication, and continuous learning has not only transformed the organizations he has touched but has also inspired countless individuals within and beyond the aerospace industry.

Navigating the complexities of a rapidly evolving landscape, Kelly has faced significant challenges, from the aftermath of the 737 Max crisis to labor negotiations and economic uncertainties.

Yet, through each obstacle, he has remained steadfast in his belief that true leadership lies in the ability to listen, adapt, and unite people toward a common vision. His approach emphasizes the importance of nurturing relationships, both within the workplace and in the broader community, reminding us all that we are stronger together. As we reflect on Kelly's legacy, it is clear that his philosophy extends far beyond the boardroom.

His dedication to social responsibility and community engagement illustrates a leader who understands that business success is intertwined with the well-being of society.

By championing initiatives that empower employees and support local communities, he has set a standard for what it means to lead responsibly in today's world. This biography captures not only the professional milestones of Kelly Ortberg but also the personal values that have guided him throughout his life.

It serves as an invitation for readers to reflect on their own journeys and consider how they can embody the principles of resilience, empathy, and innovation in their pursuits. Thank you for embarking on this odyssey with us, exploring the life and legacy of Kelly Ortberg.

Your engagement with this story is a testament to your interest in leadership, resilience, and the profound impact one individual can have on an industry and the world. May Kelly's experiences inspire you to strive for excellence, embrace challenges, and lead with purpose in your own endeavors. As we look to the future, let us all carry forward the lessons learned from his journey, honoring the values that define us and the connections that enrich our lives.

www.ingramcontent.com/pod-product-compliance
Lightning Source LLC
Chambersburg PA
CBHW071053240526
45471CB00015B/1800